# Whales & Dolphins

## KAIKOURA NEW ZEALAND

BARBARA TODD

Revised edition 2007

We need another and a wiser and perhaps a more mystical concept of animals. Remote from universal nature, and living by complicated artifice, man in civilization surveys the creatures through the glass of his knowledge and sees thereby a feather magnified and the whole image in distortion. We patronize them for their incompleteness, for their tragic fate of having taken form so far below ourselves. And therein we err, and greatly err. For the animal shall not be measured by man. In a world older and more complete than ours they move finished and complete, gifted with extensions of the senses we have lost or never attained, living by voices we shall never hear. They are not brethren; they are not underlings; they are other nations, caught with ourselves in the net of life and time, fellow prisoners of the splendour and travail of the earth.

HENRY BESTON — "The Outermost House"

# CONTENTS

*Spyhopping Orca*

*This book is dedicated to the memory of Dr Michael Bigg whose work with Orca in the United States Pacific Northwest and British Columbia helped to pioneer non-invasive cetacean research techniques. Mike's generous and sharing nature as well as his personal commitment to raising the level of understanding and appreciation of other life forms made him a unique person in the scientific community. The whales and all who knew him have lost a special friend.*

# INTRODUCTION

We share our planet with around 84 species of whales and dolphins. In New Zealand, there have been confirmed sightings of 37 of those species and in Kaikoura at least 18 species have been positively identified, a phenomenon which would be hard to duplicate anywhere in the world. Additionally, many other marine inhabitants are found along the Kaikoura coast, including numerous fish and shark species, the delicious crayfish or rock lobster after which Kaikoura was named, a colony of fur seals along the rocky shoreline, and a large variety of seabirds that can be observed skimming the ocean's surface in search of fish or krill. Many visitors come to Kaikoura to observe this marine life in a wonderful natural setting, and while they come to enjoy the total atmosphere the primary drawcard for most is undoubtedly the whales and dolphins. Its leads one to ask Why? What reason is there for this apparent empathy between man and his fellow mammals? Perhaps part of the answer lies in the words "fellow mammals" for we must remember that although they have adapted to a marine environment, whales and dolphins are mammals, like us, and must breathe air to live.

Whales and dolphins are also like us in many other respects. They possess a highly evolved brain, live in complex societies and have developed languages which include dialects and even songs. They suckle their young and form strong family bonds, and exhibit a delightful sense of humour, supposedly a sign of higher intelligence and an attribute that some of mankind no longer possess. Unlike ours however, their lives have an orderliness and simplicity, and with these a sense of peace and tranquillity that many industrialised societies have lost, or at least temporarily misplaced. Perhaps watching whales and dolphins reawakens the consciousness of our natural world – a world without artifice or pretence… a world we used to know.

Visitors arrive in Kaikoura to absorb nature, to delight in the antics of a dolphin as it leaps and cavorts upon the sea, and to

watch in wonder as a 40 tonne whale lifts his tail, and, barely creating a ripple, dives to the cold depths below in search of food. They come to ask questions, and to gather facts, which will increase their knowledge and understanding of these fellow inhabitants on our planet. This book is an attempt to answer some of those questions. But it is more than just a book of facts – it is also an attempt to share the sense of awe and appreciation that one experiences when viewing Kaikoura's marine life. In our modern age the link between man and nature has become increasingly fragile. Hopefully, with increased awareness, knowledge and respect that link will become stronger and we will be reminded that life itself, and not just human life, is the ultimate miracle upon this earth.

# KAIKOURA'S HUMAN HISTORY

## Maori History

The first humans to reach Kaikoura did so by way of the sea. One of the early Maori legends tells of the Polynesian hero Maui who "fished up" the North Island from his "canoe", the South Island. The North Island "fish" resisted so strongly that Maui had to place his foot against the thwart of his canoe for leverage. The "thwart" was the Kaikoura Peninsula. This legend is interpreted as describing part of the early discovery of New Zealand, thought to have occurred when Maui navigated his canoe through Foveaux Strait and then north along the Otago and Canterbury coast until he reached Kaikoura. It is speculated that upon landing in Kaikoura, Maui climbed the Kaikoura Peninsula for a good look around and spotted the North Island for the first time, "fishing it up" in mythological language

Around 950 A.D. another Maori explorer arrived in New Zealand by canoe. Rakaihautu found the North Island already inhabited and travelled south for further exploration, instructing his son, Rakaihouia to sail along the east coast while he explored inland. Rakaihouia landed at Kaikoura to replenish his food supplies and collected gulls' eggs along the cliffs of the peninsula. He gave the peninsula the name Te Whata-Kai-a-Te-Rakaihouia, the "food storehouse of Te Rakaihouia".

The name Kaikoura is credited to a later voyager, Tamatea-Pokai-Whenua, who is thought to have arrived in New Zealand in approximately 1450 A.D. After Tamatea's wives fled from him he gave chase, landing along the way at Kaikoura where he cooked a feast of crayfish. He called the site of his feast Kai (food) Koura (crayfish), that name surviving to the present day. Tamatea's wives, by the way, were never recovered. The legend goes that one died in Milford Sound and was turned to greenstone. Tamatea's tears streaked the stone leaving marks that are found today in Milford greenstone. His other two wives also died further up the West Coast, and they too were turned to greenstone. Three types of New Zealand greenstone are named after Tamatea's wives, Tangiwai, Kahurangi, and Kawakawa.

The descendants of today's local Maori, the Ngai-Tahu, arrived in Kaikoura during the 17th century, and ultimately assimilated with the earlier occupants, the Ngati-Mamoe. The Ngai-Tahu prospered in Kaikoura for almost two centuries before suffering heavy losses to a northern tribe, the Ngati-Toa, led by Te Rauparaha. By the early 1840s when whalers began to consider Kaikoura as a possible site for shore whaling, only a small but strongly united tribe of 40-60 Ngai-Tahu remained under their chief, Kaikoura Whakatau.

## Pakeha (European) History

In mid-February 1770, Captain Cook, on his voyage of discovery around the Islands of New Zealand, recorded the sighting of Kaikoura in his log book, calling the place "Looker's On". There is little evidence of other European visitation during the next 70 years although sealers undoubtedly made an impact on these shores, decimating the fur seal colony as they did everywhere else in New Zealand where these creatures were found. Interest in Kaikoura increased in the early 1840s with the possibility of colonization by the "New Zealand Company of England". These plans however were set aside as reports of difficult land access and poor sea shelter reached England. Whalers on the other hand were not put off by these drawbacks. Their catches from other New Zealand shore stations were declining and Kaikoura appeared to them a promising area for the establishment of a new operation. Plans for the first station were laid in the spring of 1842 and Kaioura's first shorebased whaling started in late April of 1843 with 40 men. By 1845 other stations were operating, employing over 100 men. But by this time shore whaling was becoming a dying trade. The Right Whale population was rapidly declining and even catches of Humpbacks were decreasing. However, shore whaling continued out of Kaikoura with a succession of owners until 1921, but subsequent seasons were never as successful as those in the early 1840s.

By the 1850s, more and more intrepid individuals were viewing Kaikoura as a place to settle and "run" sheep. Then, in late February of 1864, 15,000 acres of land was auctioned off in smaller parcels for family occupation. The half acre town sections sold for £10 each, and suburban sections of 40-50 acres were selling for £2 per acre. By 1866 Kaikoura was a bustling settlement and the first race meeting was held at South Bay on Boxing Day.

Until 1963 Sperm Whales formed an insignificant part of the catch taken by New Zealand shore-based whalers. Sperm Whales were exploited, however, by other pelagic whaling nations, although their numbers never dropped as low as for the Right and Humpback Whales. In 1963 and 1964, however, New Zealand whalers made a concentrated effort to catch Sperm Whales, operating mainly between the eastern end of the Cook Strait and around Kaikoura. During that time a total of 262 whales were killed with 248 of those being Sperm Whales. The dropping oil prices in 1964 led to the end of the killing and to the closing down of New Zealand's last whaling operations. The Marine Mammal Protection Act was passed in 1978, finally affording total protection to New Zealand's whales, dolphins and seals.

*"And when the last great whale died, no sigh was heard upon the land, but in the heaving of the tide, with every throb, oceans cried, and cursed the ways of modern man …"* ( Unknown )
Colin Monteath-photo

# KAIKOURA TODAY

Kaikoura, the coastal village where the "mountains meet the sea", has always been known for its spectacular scenery. Until the mid 80's the mainstays of the economy were farming, fishing, and the railway, and most people passed fleetingly through the area, their thoughts on either the views or obtaining a feed of crayfish. However an increased awareness of Kaikoura's natural wonders, particularly its marine life, has turned the former trickle of tourism into a steady stream, and today Kaikoura bustles with activity and optimism.

*Whales have once again become an important factor in the Kaikoura economy; but happily today's excursions are not to obliterate the whales but to observe them.*

*A wide variety of species are commercially and privately fished at Kaikoura including cod, butterfish, groper, kahawai, tarakihi, shark, paua, and crayfish. (The whales and dolphins love the fishing too!!!)*

*For many of us whales have become symbolic of a level of consciousness of the natural world that we are beginning to re-explore in our own inner nature, a consciousness that has not been totally lost in today's modern world, but has merely been sleeping in our own interior wilderness.*

# WHY KAIKOURA?

Kaikoura has now earned the reputation of being New Zealand's "Natural Marineland". To understand why Kaikoura has such abundant and diverse marine life, we must first examine the total oceanic community, which is interconnected by a huge food web.

The food chain begins with decomposed organic matter which has drifted to the ocean's floor. This matter is carried to the surface of the sea by the ocean's upwellings and is mixed together by the ocean currents. The decomposed organic matter is a nutrient for microscopic plants known as phytoplankton. Phytoplankton include diatoms and other forms of algae and live near the sea's surface where they synthesize light into energy.

Plankton has limited ability to swim and is mainly carried by the currents. The drifting phytoplankton are consumed by small animals called zooplankton which consist of many creatures, among them tiny crabs, small molluscs called copepods, sea worms, and the "shrimplike" krill. The zooplankton in turn are consumed by small schooling fish, squid, basking sharks, penguins and other seabirds, some seals, and baleen whales. The smaller fish and squid are eaten by larger fish and shark species, and various sizes of fish and squid are consumed by seals and the toothed whales and dolphins. It's basically a gastronomical pecking order! This continuing food chain is fuelled by the ocean's currents which cause the upwellings which bring nutrients to the surface.

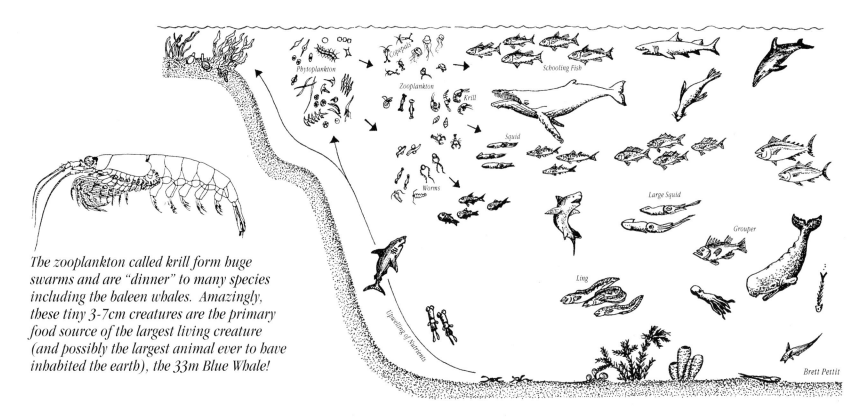

*The zooplankton called krill form huge swarms and are "dinner" to many species including the baleen whales. Amazingly, these tiny 3-7cm creatures are the primary food source of the largest living creature (and possibly the largest animal ever to have inhabited the earth), the 33m Blue Whale!*

Brett Pettit

10

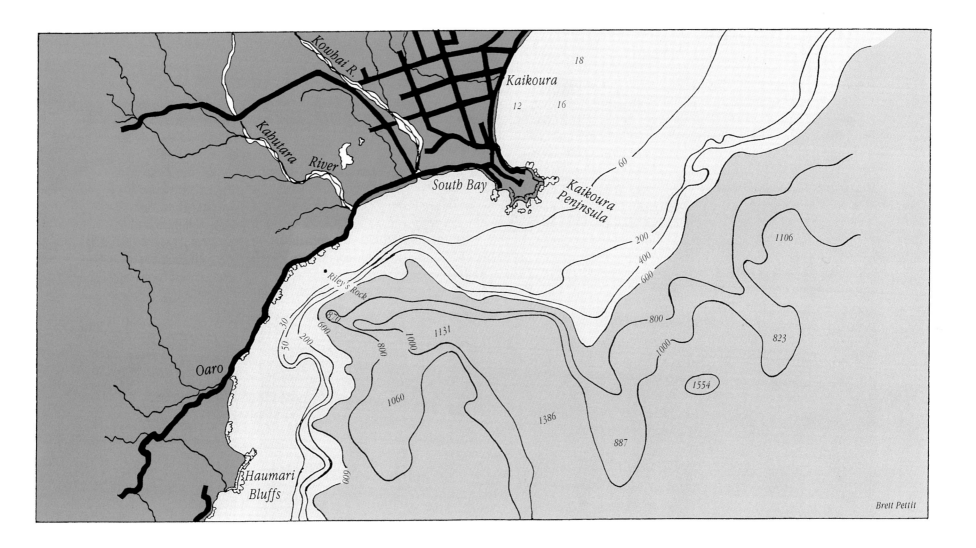

Brett Pettit

Kaikoura is an ideal location to support the huge food web that sustains life in the sea. A cold north-moving coastal current is met by a warm offshore southward-moving current, and a fairly constant upwelling is maintained from the convergence of these currents. Deep underwater canyons in the Hikurangi Trench run through the Kaikoura area with depths of 870m occurring only a kilometre or so offshore. A little further out, the depths of these canyons exceed 1600m in some places, and they contain many of the deepwater fish and squid species so favoured by Sperm Whales. These combined factors create a wonderful banquet for Kaikoura's shellfish, fish, sharks, birds, seals, dolphins, and whales. It's almost an "All you can eat" buffet, and makes it easy to understand why the whales and dolphins find Kaikoura such an attractive place to visit or live. But what about the whales and dolphins themselves?…. How did they ever come to be?

# THE EVOLUTION OF WHALES AND DOLPHINS

The exact origin of cetaceans remains a mystery. Evidence suggests their early ancestors were four-legged terrestrial mammals who began adapting to life in the sea at least 50 million years ago. These creatures lived on the edge of swamps and other waterways, and as they spent more and more time in the water physical modifications began to occur. Their bodies became sleek and streamlined as forelimbs became flippers, hindlimbs eventually vanished, and powerful tailflukes appeared to act as propulsion. Insulating hair was replaced by blubber and their nostrils gradually migrated to the top of their heads for more efficient breathing. Ultimately such adaptations allowed these creatures to abandon the land totally for a marine environment, and they became the whales and dolphins that we know today.

Three groups of cetaceans developed from these early ancestors. The first group, the Archaeoceti, were mainly long, serpentine creatures who became extinct approximately 20 million years ago. The second group evolved into today's toothed whales, the Odontoceti. And the third group lost their teeth and developed filtering fringes known as baleen. These are the Mysticeti, the "moustached" or baleen whales. Whales, dolphins, and porpoises are further divided into families and species. There are approximately 84 species of living whales, dolphins and porpoises. We say approximately as there continues to be debate in the scientific community about the classification of some species. All these creatures are collectively called "cetaceans" which comes from the Latin word "cetus" meaning "whale".

*"Diviner than the dolphin is nothing yet created; for indeed they were aforetime men and lived in cities along with mortals, but by the devising of Dionysos they exchanged the land for the sea and put on the form of fishes." Oppian (Third Century Greek poet)*

# MATA-MATA

A well-known chief and brave warrior of the tribe of Ngati Kuri, *Rakaitauneke, lived at Tahuna Torea (Goose Bay). In the sea directly opposite to his dwelling lived Mata-mata, the whale, whose sole duty was to do Te Rakaitauneke's bidding, to serve all his needs, and to guard him against harm when he was able. Wherever Te Rakai went Mata-mata went too. When he went to Takahanga, Mata-mata could be seen blowing outside the garden of memories, as close to shore as he could possibly get. Rakaitauneke's love of his whale was as great as Mata-mata's love for him.*

*After Rakai's death Mata-mata was not seen along the Kaikoura Coast for many months and some people believed that he had gone away and died of sorrow for his beloved Rakai. But there were those who remembered Rakai's prediction that after his death Mata-mata would only return when a relation of Rakai was facing imminent death or danger.*

*And so it was that while working in the garden with his wife, a close relation of Rakaitauneke remarked on the noise that a whale was making offshore from where they were working.*

*"Don't worry about the noises he is making, husband, just be grateful that we can share this day together because that whale is Rakai's whale, Mata-mata, and he is here to tell me that my time is about to end." He looked at her sadly, "perhaps the day could be better shared if we were to sit and relive our lives and remember those happy times."*

*"You have the rest of your life to do that my husband. It is hard enough that I have to leave you. Let us continue our work." He sighed sadly. "My time will be shortened by your passing my wife and already I look forward to joining you." So it was that she died late the next day fulfilling Rakai's prediction.*

*Many of the descendants of Rakai when faced with danger on the high seas have been saved by the timely intervention of a whale. And so it is to this day that the descendants of Rakaitauneke look with a certain amount of trepidation at the seas but those, like myself, who work on it face it with a lot of confidence.*

*- as told by Bill Solomon, Ngai-Tahu, Kaikoura.*

# ODONTOCETI: THE TOOTHED WHALES AND DOLPHINS

There are 71 species of toothed whales and dolphins, ranging from the largest, the 16-20m Sperm Whale, to the smallest, the tiny 1.4m Hector's Dolphin. Kaikoura is unique, for both the Sperm Whale and the Hector's Dolphin reside along its coast and are observed almost daily. Odontocetes use their teeth to seize prey such as squid or fish, which is then generally swallowed whole. In order to help locate and catch their food, the toothed whales and dolphins, like bats, have developed a sonar or echolocation system, in which sounds are emitted from their heads and reflect off solid objects. The returning echo is interpreted by the whales and dolphins and enables them to determine the size, composition, distance, and direction of the object they have focused on.

If this all seems terribly confusing, think of a whale's head as a type of acoustic lens. As a camera lens collects reflected light from an object and forms an image on film, so a whale's echolocation system gathers reflected sound which forms an image on its brain - and they don't even have to go to a "one hour" lab!! It's a truly amazing system and one that we might well envy. We have only developed machines that utilize sonar capabilities in recent years, and these systems are still far less efficient than those of the whales and dolphins. In addition to their echolocation system, most toothed whales and dolphins have developed the ability to communicate by using clicks, whistles, squeaks and squeals, and some, such as Orca, have even developed dialects which differ from pod to pod. Many toothed whales and dolphins utilize these communication skills to help catch their prey, and form social groups to hunt cooperatively.

Odontocetes feed in diverse surroundings with some of them utilizing the ocean's deepest depths, usually far offshore, while others feed nearer to the surface and closer inshore. Because of Kaikoura's diverse depth range many species are observed which would not normally inhabit the same area. We find deepwater feeders such as the Sperm and Pilot Whales sharing the same waters as oceanic near-surface feeders and inshore feeders such as Orca, and the Common, Dusky, and Hector's Dolphin. A total of at least ten different Odontocete species have been positively identified on more than one occasion off the Kaikoura coast.

*All toothed whales and dolphins have a single external blowhole.*

14

"They say the sea is cold, but the sea contains the hottest blood of all, and the
wildest, the most urgent. All the whales in the wider deeps, hot are they as they
urge on and on …the Right Whale, the Sperm Whale, the Humpback, the
Killer … there they blow, there they blow, hot wild white breath out of the seas."
D.H. Lawrence (Whales Weep Not)

# THE SPERM WHALE

Try to imagine for a few moments how it must feel being the largest toothed whale, a whale which is different in many respects from its other cetacean cousins. To begin with, you have an enormous head which is up to 1/3 of your total body length. Inside your head you store a large quantity of very fine oil which made you a sought after prize by the whalers, oil which in earlier times was thought to be stored sperm, hence your name, Sperm Whale. If you're a male, you reach a maximum of 16-20m, but if you're a female you are much smaller, reaching maximum lengths of 12m. While resting and reoxygenating on the ocean's surface, you display very little of your huge bulk, resembling more, as one ten year old observer noted, "a smoking log". It's not until you lift your massive tail flukes to begin a dive in search of food that your might is truly exhibited. And it's during that dive that you accomplish some amazing feats, feats possibly not equalled by any other living mammal.

As you descend into the dark icy sea, you are travelling at approximately 3 knots, and the depths you are capable of reaching are phenomenal. You have been tracked on man's sonar system to over 2000m under the sea, and because of certain bottom dwelling shark species which have been found in your stomach, there is a suggestion that you are able to dive to depths in excess of 3000m! The pressure at such depths is incredible – how do you survive? To begin with, your body contains a higher percentage of blood than humans, making it more liquid and therefore less easily compressed. You have a flexible ribcage and as you dive your chest partially caves in as your lungs collapse forcing any remaining air into the nasal passages leading to the

*Above: A piece of squid floats to the ocean's surface after being chomped off by a feeding whale. The remains of 12m giant squid have been found in the stomachs of Sperm Whales.*

*Left: On the surface, a Sperm Whale displays approximately 2/3 of its overall length from the front of its body to the dorsal hump. The slight indentation 1/3 of the way back is the end of the whale's massive oil-filled head.*

blowhole. Over 40% of your total oxygen supply is stored in myoglobin-rich muscle tissue and this is what you survive on when submerged. To help conserve oxygen, a condition known as "bradycardia" occurs, a slowing of your heart rate and a redistribution of oxygen-rich blood to vital organs such as the brain and heart. You have been known to remain submerged for at least two hours making you one of the breath-holding champions of the sea.

When you reach your chosen depth, the search for food begins. Your prey is primarily squid, although you often enjoy a bit of variety and feed on other deep-dwelling fish species such as ling, cod, groper, and shark. Occasionally you even make a bit of a blunder, for strange items such as glass fishing balls, gumboots, coconuts, and even an oil drum have been found in your stomach. While searching for food in the dark depths, you use your sonar or echolocation system as your eyes, your strongest sense at these depths. You continuously send out sound beams, and as those beams reflect off solid objects they bounce back to be collected, perhaps by your oil-filled jawbone, and transmitted into the huge spermaceti case and to your brain. Scientists are still somewhat confused about how your echolocation actually works and about how you catch your food. Do you chase it down, or is it attracted to the white area which surrounds your lower jaw? Or do you, as some scientists suggest, use your sonar as a type of weapon, sending out powerful bursts of sound that stun and debilitate your prey making it "easy pickings". Whatever method is used, when you catch your food, you swallow it whole, except for occasional bits such as squid tentacles, which may get chomped off by the 18-25 pairs of large teeth which are only in your lower jaw.

When it's time to replenish your oxygen supply, you emerge from the depths with a big "whoosh" of air. Occasionally you resurface with a giant leap (or "breach") which sometimes totally clears your body from the water. Imagine propelling up to 50 tonnes clear of the water! While on the surface you normally lie quietly for 10-12 minutes, breathing 3-4 times per minute. Once your system is recharged, you take one last big breath and then lifting your enormous tail you gracefully descend into the deep and the feeding cycle begins again.

*"...when he is about to plunge into the depths his entire flukes with at least 30 ft. of his body are tossed erect in the air and so remain vibrating a moment till they downward shoot. This "peaking" of the whale's tail flukes is perhaps the greatest sight to be seen in all animated nature." Herman Melville (Moby Dick)*

# PHYSICAL AND SOCIAL CHARACTERISTICS OF THE SPERM WHALE

Sperm Whales are the most sexually dimorphic of all the cetaceans. Males have reached lengths of 20m, while the maximum length for females is 12m. Both sexes are normally smaller with mature males averaging 17-18m and mature females 10-11m. Their weights vary from 32,000-45,000kg (35-50 tonnes) for the males and up to 20,000kg (22 tonnes) for the "petite" females. Calves average 3.1-4.3m at birth and weigh around 900kg. Imagine giving birth to a 4 metre, 1 tonne baby after a gestation period 14-16 months! It's believed Sperm Whales have a life span similar to that of humans, 50-70 years.

The migratory and social behaviour of Sperm Whales also varies greatly. Females, calves, and sexually immature whales spend the majority of their lives in tropical and temperate waters, ranging from the equator to latitude 45° moving into the lower latitudes during the winter months. Calves in the Southern Hemisphere are thought to be born from late spring into summer, indicating that most of the breeding probably occurs from late winter into the spring. From spring to autumn mature bulls migrate and feed in the higher latitudes with the largest, oldest, and often solitary bulls travelling to the icy polar waters of the Southern Ocean. The females, calves, and immature whales also move into cooler waters during those months, but they do not normally travel into latitudes greater than 40- 45°. After a long summer season of feeding, the whales return to their breeding grounds. Research suggests that females breed every 4-6 years and that they nurse their young for approximately 2 years, although whales as old as 10-13 years have been found with milk in their stomachs indicating they "sneak" a drink on occasion. There is evidence that communal babysitting and suckling takes place.

Like all youngsters as the young whales mature they seek independence, and there is an indication that they form groups with other individuals ranging from 4-15 years of age. These "immature" groups of whales remain in the same areas as the females but with increasing independence and autonomy as they get older.

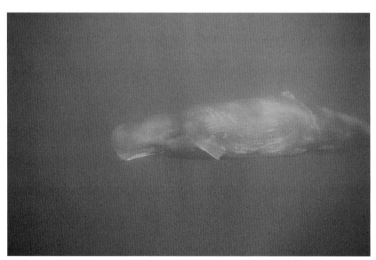

*A female Sperm Whale off the Galapagos Islands clearly demonstrates the white area surrounding the lower jaw, which may help to attract squid. Mark Jones-photo*

*The Sperm Whale's boxlike head contains the largest brain of any creature that has ever lived.*

As the males become sexually mature (usually between 14-20 years of age), they leave the female and immature groups, either of their own volition or perhaps by "invitation".

The acceptable breeding age for male Sperm Whales appears to be at least 25 years. It was once thought that these breeding bulls controlled harems and fought bitter battles over "their women". Studies today indicate that this is not the case and instead two or three bulls may travel from one group of females to another seeking out "willing companionship". Nonetheless, these breeding bulls would not be terribly excited over younger competition and some aggression undoubtedly occurs on the breeding grounds.

But getting back to our sexually maturing, but not yet sexually acceptable, young whales. They are faced with the age-old dilemma of what to do with themselves for 5-10 years and so they do what teenagers have done for centuries, they form "bachelor" groups and "hang out" together. The youngest whales form the largest groups and as they become older the groups appear to get smaller, ultimately ranging between 3 and 15 individuals. It does not appear that these young "hoons" ("hoon" is a New Zealand term for an unruly male teenager) migrate into the breeding areas in winter; instead they remain in their feeding territories getting bigger, older and wiser. We believe that the majority of the whales that we regularly view in Kaikoura are these young "hoons" ranging in age from 15-25 years.

*The Sperm Whale's blowhole is asymmetrically located at the front left-hand side of it's head, causing it's blow to angle forward and to the left.*

# WHAT WE'VE LEARNED ABOUT KAIKOURA'S SPERM WHALES

While working in Kaikoura we accumulated a large amount of local data and information about its marine life, particularly the Sperm Whales. I say "we" for many people have gathered and are still gathering information which is communicated amongst us. We all know that it is a privilege for us to gain some glimpses of the marine world, a world which is in many respects completely foreign to us. As we've watched the birds and mammals day after day some sensations have been shared by all of us, including the continuing sense of wonder and awe that we experience as we observe these creatures living, playing, and feeding, in their natural environment. It would be impossible in this small publication to divulge all we've learned, so I will attempt to give a brief synopsis of our information and experiences.

We knew when we started Nature Watch (the first whale watching operation in Kaikoura) that Sperm Whales used the Kaikoura area as a feeding ground. At that stage some of our unanswered questions were:

1. How many months of the year were the whales present?
2. Were the same individuals present all the time?
3. Was it possible for us to positively identify each individual?
4. What were average dive and surface times, and did the whales have preferred feeding areas?

To date we have answered some of these questions, but in the process come up with a whole new series of questions awaiting answers.

Our first problem of course is to find the whales. It's a big ocean out there and the whales, as we've already mentioned, somewhat resemble a log when on the surface. If you've spent any time at sea, you know how hard it is to see logs! But these are "smoking" logs and therein lies the secret. The "smoke" is the whale's blow, or spout, which appears every time the whale exhales. The blow is made up of warm, moist air which condenses upon entering the outside atmosphere and becomes very wet. I should add that

whale breath is also very smelly - they definitely don't use mouthwash! Sperm Whales blow 3-4 times a minute while on the surface and that blow is usually our first visual contact. We have another way of tracking the whales, however, and that is by sound. While underwater, the whales continuously emit their sonar clicks, and by means of an underwater listening device called a hydrophone we can hear those clicks and know if a whale is close by. Hydrophones vary in their sensitivity; we found that a range of approximately 1.5km worked well for us, and if we could hear the whales on our hydrophones, we could usually rely on seeing them when they surfaced. With this combination of listening and looking it meant that we could almost always find whales if they were in our area.

After four years monitoring we discovered that whales are present all year round, although some months have higher concentrations of animals than others. During the autumn and winter months (April to August), large runs of spawning hapuka (groper) come into Kaikoura's waters. The whales are delighted, for it means an added variation to their staple diet of squid, and the occasional ling or shark. In the autumn and winter months some whales move much closer inshore, and certain individuals appear to have favoured feeding areas. We often saw the same "friends" almost daily: "Elephant Ears", "Groove", "Droopy Flukes", "White Spot", and of course "Hoon" (more about him later!). These whales, as well as others, "worked" these areas seasonally for at least four years. Interestingly we have observed "White Spot", "Droopy Flukes" and some other known individuals at other times during the year, but we never saw "Hoon" in the summer months. When the groper run ends (from late July to late August), the whales spread out and feed further offshore. Many winter whales disappear and do not show up again until the following autumn and winter season, while others reappear throughout the summer months, along with individuals that we do not see in winter.

We are often asked, "how many whales are in the Kaikoura area"? It is a difficult question to answer for the boats have a limited range and there are undoubtedly whales feeding 30km or more offshore which we never encounter. Research has now confirmed our original estimation that between 80-120 whales move in and out of the main Kaikoura whale watching range. There are approximately 25 known whales which are seen frequently, another 25 or so which are seen less frequently, and various other individuals which are seen on occasion. On any given day only one or two whales may be observed and they are often sighted singly. This makes sense if you think about it, for the whales are utilizing the area as a feeding ground, and like any clever fisherman, they are not all going to fish in the same spot. The depths the whales feed in vary between 600 and 1600m. The average dive time for the whales is 40-45 minutes, but the duration of dives varies, with shorter dive times observed in winter when the whales have added groper to their diet, and longer times in summer when the whales feed almost exclusively on squid. Our longest recorded dive time of a single individual was one hour and forty seven minutes – it seemed like forever!! Interestingly, the surface time for the whales does not vary a great deal, and normally averages 10-12 minutes. Some whales have shorter times, and we have occasionally sat with whales at the surface for 30-40 minutes. We definitely notice distinctive routines in the movement of the whales in the winter, and certain individuals have very preferred areas in which they feed. In summer, these routines tend to be less obvious, although there are still some distinctive patterns.

On a few occasions during the summer months we have been delighted to discover a large group of up to 50 female Sperm Whales in the area. These sightings are rare occurrences however, which is to be expected, as Kaikoura is located at latitude $42.5°$, close to the limit of the female's normal range. When sighted the females only remained in the area for one day and were reasonably shy of the boats. Their tiny size compared to the males made them seem like "toy whales". We have also witnessed groups of over 30 very large males, whales which we believe were mature breeding bulls and who were perhaps migrating into the higher latitudes. These whales averaged 15m plus – they were humongous! At the other end of the scale small males in loosely aggregated groups have been sighted, all under 12m and possibly "new" bachelors. The majority of whales which we observe appear to average 13.5-15m and we believe that these are bachelor whales, that are sexually mature but not yet sexually acceptable as breeding partners. Even though the Kaikoura whales are normally observed singly, there are often 4-5 whales "working" an area of maybe 2-5 km square, and when this occurs there is a definite coordination in their dive times, with the whales surfacing and sounding within minutes of each other.

There is still much to learn about the social behaviour and aggregations of Sperm Whales. They spend the vast majority of their lives "on another planet", the deep and dark depths of the sea. It is only in recent years that studies of living Sperm Whales as well as other whales and dolphins have taken place. Our knowledge has improved but is still incomplete, for studying nature's creatures is like trying to put together a 1,000 piece jigsaw puzzle – each piece adds a new dimension, but often changes the total conception of what we thought the picture was going to be. Many questions remain unanswered, though that is not necessarily a bad thing. Perhaps, one question we should ask is: "How much does man really need to know?" We humans always seem to feel that we must take other forms of life and try to give them order - our order. We factualize, theorize, categorize and computerize, and somehow that gives justification for existence. But facts do not necessarily create total understanding, respect, or appreciation. And what we often lose is a sense of mystery and awe. Whales and dolphins live, breathe, sing, play, make love, rear their young, grow old, and die... they exist, they are. Perhaps that is all we really need to know.

# VOCALIZATIONS

We, like many others, had listened to the haunting songs of the Humpback Whales and had heard the enchanting "squeaks and squeals" of dolphins; but the vocalizations of Sperm Whales were something we knew little about. We knew that Sperm Whales do not use whistles and calls as part of their vocabulary, but instead utilize a series of clicks which range from around 200 Hz to 32 kHz, with the dominant frequency being around 5 kHz. To most humans, Sperm Whale clicks appear very repetitious and probably somewhat boring. A single whale sounds like a person slowly typing with one finger, or a very lazy carpenter hammering a nail; a group of whales sounds more like a typewriter pool or a gang of industrious carpenters. The more we listened to those "boring" clicks, the more fascinated we became. We knew the sounds we heard were being used to navigate, find food and communicate, and we started trying to visualize what events were actually occurring hundreds, and also thousands, of feet below us as we listened on our hydrophone.

The sounds we hear most often appear to be scanning or searching clicks. The whale sends out a steady pulse of sound at a little less than a second apart. Suddenly those pulses become faster, closer and closer and closer together until they trail off to a Bzzzzzz, normally followed by a period of silence that lasts from 10 to 40 seconds. We call this the "click, click, click, gulp" theory and visualize the whales below searching for their food with the steady pulse, locating and closing in on their prey as the clicks become faster, and then silence - snack time! Following the silence, the clicks once again assume their steady rate. Another interesting click that we hear is a slow, steady pulse which occurs from 4-6 seconds apart. These clicks are extremely intense – in fact I've had to remove my earphones at times as the sound reverberated through my head. About 95% of the time when we hear "slow clicks" a whale will surface within 2-5 minutes. Often the whale will continue to "slow click" briefly after he surfaces. We call these clicks "surface clicks" and surmise that the whale may be just under the surface of the sea, acoustically checking out the environment into which he is about to emerge.

Other clicks we hear are harder to describe and harder to interpret. They are possibly communication clicks as they have a varied cadence, unlike the steady pulse click that suggests pure echolocation. Sometimes it almost sounds like Morse Code and we often hear these varied patterns when the whales are travelling underwater. Perhaps they are chatting away to each other as they search for a different area with a better food supply? We also often hear these patterns near the end of a dive – perhaps the whales communicating their intention of surfacing? For the moment however, our interpretation of these sounds is definitely speculative, with most of the mystery still to be solved.

One of the more fascinating behaviours we observed was the "tail first surfacing", normally performed by our friend "Hoon". One of the interesting aspects of this behaviour was the vocalization which occurred on every occasion. We heard the slow surface clicks and then a long "click train", a series of loud closely spaced clicks which trailed off as Hoon's tail came shooting out of the water. On one occasion we heard the same sequence prior to a breach by Hoon, and since then we observed Groove and White Spot emerge tail first after emitting the same "click train". Perhaps they were jealous of the attention Hoon received with his trick!

# PHOTO IDENTIFICATION

One of the most important breakthroughs in cetacean research was the discovery that whales and dolphins could be individually identified from natural markings on their bodies. Some of the earliest work was carried out in the northwest United States and Canada where researchers, by learning to individually identify hundreds of Orca, gained vast amounts of new knowledge regarding their social patterns, family histories, vocalizations, and the extent of their home ranges. These identification techniques have been applied to other species of whales and dolphins, giving scientists and laypersons alike new insights into migration patterns, social behaviour and population numbers.

Different species of whales and dolphins are identified in different ways. For example, Orca are identified by the greyish-white saddle patch which is behind their dorsal fin. Humpbacks are identified by the pigmentation patterns on the underside of their tail flukes. Right Whales are identified from the unique callosity pattern on each whales head. Additionally, markings, such as nicks, cuts and scars are also used.

For a time, Sperm Whales posed a real problem in terms of individual recognition. Their dorsal humps were distinctive in some individuals but in others were hopelessly similar; they have no saddle patches; and the undersides of their flukes are black, black, black. Help! Happily a variation surfaced, or perhaps we should say "sounded", for researchers discovered as they watched Sperm Whales that their tail flukes all have a distinctive scalloping pattern along the trailing edge which is exhibited as they dive. At last there was a means of telling the "smoking logs" apart!

Researchers in the Galapagos, the Indian Ocean, and the Azores are using their cameras to learn more about "Sperms" and we in Kaikoura are doing the same. To date at least 119 whales have been individually identified in Kaikoura, information that has been indispensable in adding to our knowledge.

*Above: "White Spot" is one of our friendly whales who made appearances in both winter and summer. He and "Hoon" spent a bit of time together - competitors or mates??*

*Left: "Elephant Ears" was named after his huge floppy tail. He has a very obvious scalloping pattern along the trailing edge of his tail flukes which makes identification an easy task.*

*"Groove" - another "favourite" we've known and who predominantly appears in winter. Like "Hoon" we watched him grow larger with each passing year – He's looking very "grown up" these days!*

*This picture shows the underside of "Groove's" tail flukes. A smooth circle is starting to form in front of the tail. The whales leave this circle behind when they dive – their "footprint".*

*"Vee" was named by Greg, a young passenger on one of our trips. As soon as we photographed a new whale, we asked our passengers to name him – some interesting names include "1991", "Schizo" and "Bonus".*

*"Droopy Flukes" a rather obvious name! "Droopy" is quite large and is observed both in winter and summer. He's sometimes quite curious and has made at least four close approaches to our boats.*

25

# "HOON", WHO WAS THAT WHALE???

In April 1988 we had been operating our Nature Watch excursions for almost two months, finding the majority of our whales between 12km and 20km offshore. We knew, however, that it was only a matter of time before we started seeing whales feeding close to the shore in conjunction with the autumn and winter runs of groper. One April day we decided to do a "coast run", stopping and listening for the sounds of any whales that might be in the area. Approximately 1.6km off Goose Bay we threw in our hydrophone and heard a booming click. "Hooray", we shouted and settled down to wait for the whale to emerge. When he surfaced, we started our engines and travelled slowly towards him, keeping in mind that he was a "new" whale and might be suspicious of our boat. We moved to within 60m and he displayed no signs of distress; we edged 10m closer and stopped. The whale took another breath, brought his head out of the water for a good look at us, then resumed his steady blowing. "Aha", we said, "this could be a good whale". Little did we know that he was to become not only our favourite friend but perhaps the most famous Sperm Whale in the world. A few trips later, an excited boatload of "Kiwis" decided that "Hoon" should be his name, as he was obviously a teenage bachelor male. From then on we saw Hoon almost daily until the end of the groper season in early August. Hoon delighted locals and tourists alike with continuing displays of somewhat atypical male sperm whale behaviour.

We tend to feel a sense of aloofness from the majority of the whales we observe, as they tolerate our presence but few display much curiosity. Not Hoon though! The first winter he spent a lot of time "looking" at us, and if we weren't close enough to him, he would just swim over to get closer to us. One day he ended up with his body under the bow of our boat, head on one side, tail on the other! Other days, he would lie alongside the boat, blowing his bad whale breath in our faces. A few times after sounding, when we announced that he was gone for an average Hoon dive of about 30 minutes, he would suddenly reappear on the other side of the boat as if to say, "you don't know everything yet"!

Then there's the tail-first surfacing trick described in the vocalization section. One of our favourite stories is of a stormy winter day when we decided to cancel our trips. It's almost impossible to see whale-blows in a whitecapped sea, but we had five "desperate" tourists who were leaving the country knowing they would probably never again have the opportunity to view a Sperm Whale. They pleaded, we gave in, but warning them that our chances of seeing anything were slim. One and a half hours later, we had heard Sperm Whales on the hydrophone, but had seen nothing, and decided it was a hopeless venture. We turned the boat towards home and two minutes later a tail shot out of the sea and remained in the air "waving" at us. It was amazing, and it was Hoon of course. The people were ecstatic, and Hoon being the performer that he was remained on the surface for over 20 minutes, about twice as long as normal, before diving down into the sea.

We saw Hoon for four seasons. He was often the first "winter whale" observed close to shore, usually arriving between late March and early April. He was commonly one of the last "winter whales" to depart between mid-August and early September. Hoon continued to captivate audiences with his antics and spectacular high-fluking dives that showed off his perfect "Robert Redford" tail. In 1991 Hoon departed from Kaikoura in mid-August. Three weeks later one of the whale watching boats discovered a dead whale floating out at sea. There were no obvious clues as to the cause of death and it was difficult to positively identify the whale but it looked as if it was Hoon. A 'karakia', a special Maori ceremony, was performed at sea for the whale. By mid-April 1992 our fears were confirmed when Hoon failed to appear. Kaikoura had lost a cherished friend.

We have come to know many whales over the past years and everyone who works in Kaikoura has a favourite. Hoon was definitely the best!!

*Hoon displays his "Robert Redford" flukes.*

# ORCA

Orca are one of the handsomest whales, and also one of the most intriguing. Until fairly recently, Orca had earned a reputation for fierceness, but that image has been negated, at least as far as humans are concerned, as we have learned more about their friendly and gentle nature. We, as well as many others, have spent countless hours around Orca in various sized boats, kayaks, and even swimming with them. They have shown curiosity and have played with us, but have never shown any sign of aggression. To date there is no confirmed report of an unprovoked attack on humans by Orca at sea.

Orca have the most varied diet of any cetacean, dining on various fish species as well as seals, seabirds, penguins, sharks, and other dolphins and whales. The name "killer whale" was derived from "whale killer", a name given to them by early whalers. There is a fascinating story of whalers and Orca who worked cooperatively together for many years around the turn of the century off the southeast coast of Australia. If Humpbacks or other whales were around, the Orca would come into Twofold Bay and leap and splash alerting the whalers of their presence. They would also assist the whalers in rounding up the other whales. In return, the whalers would allow the Orca to feed on the dead whales for a time before hauling them in for processing. One of the male Orca was named "Old Tom", and his skeleton is displayed in the Eden Whaling Museum.

*Orca are one of the fastest of the whales and dolphins, reaching speeds of up to 25 knots.*

28

Comparison between Orca and wolves is an apt one, for both highly intelligent species are top predators in their respective worlds and utilize cooperative and coordinated hunting techniques. Additionally both species live in matriarchal societies with highly complex social structures. Resident Orca, which have been studied since 1972 in the northwest United States and Canada, appear to remain in family groups or "pods" throughout their lifetime, with only rare instances being recorded of members leaving the natal pod. Orca utilize two types of sound; sonar or echolocation clicks are used to assist in navigation and in locating and catching prey, and a variety of "squeals" and "whistles" are used in communication. Research has shown that each pod has a specific set of calls known as the pod's "dialect" and these dialects are so different that each pod can be identified by merely listening to their calls. Orca have a lifespan similar to that of humans. Females appear to live longer, potentially reaching 80 plus years, while males appear to have a 50 year average. Females have a gestation of 15-16 months and nurse their calves for up to two years. Orca are slow breeders averaging 5-7 years between calves and only producing 4-6 offspring in a lifetime, generally over a period of 20-25 years. Although worldwide distribution of the species ranges from the equatorial to the polar regions, Orca do not appear to undertake extended migrations, but instead seem to have a "home" range which is undoubtedly governed by food availability and, in polar regions, by the movement of the sea ice.

*Orca are individually identified by the greyish, white saddle patch which is located behind their dorsal fin. This saddle patch has a different configuration or pattern for each individual whale.*

At least three separate pods of Orca pass through Kaikoura. They are normally sighted more frequently in summer and often follow the same route as they pass through. Sometimes they remain in the area for a day or two before moving on. And sometimes they stop for a snack... fish, sharks or Dusky Dolphins.

In a rather bizarre turn of events, we once observed Orca travelling south through Kaikoura and harassing Duskies as they went. Then three days later the same group of Orca arrived travelling north, and on this day they swam and surfed with a large group of Duskies with the dolphins showing no apparent concern and even riding the bow wave created by the Orca.

*117 individual Orca have been identified by researchers in New Zealand. By comparing photographs, researchers have learned that the Orca are travelling between the North and South Island.*

29

We played with a small group of Orca for over three hours one day. They repeatedly approached our 6m inflatable blowing their smelly breath in our faces and "spyhopping" for a better look into our boat. At one point, four Orca approached the boat and rested their heads "all in a row" on the side of our pontoon. One Orca, in particular, was intrigued by our motor and kept turning upside down under our prop for a "whirlpool bath". I have spent many exciting hours with Orca but this encounter was the most memorable. Finally we had to depart and I leaned over the bow calling goodbye. Suddenly a huge male turned, swam back to the boat and gently placed his towering dorsal fin into my outstretched hand... a final farewell from the whale called "Killer".

*Orca always bring a kind of magic when they appear and almost more than any other mammal create excited choruses of "oohs" and "aahs".*

*Male Orca may reach lengths of 9.5m and weigh 10 tonnes. Females may reach lengths of 7.5m and weights of 7-8 tonnes. The male's outstanding dorsal fin can grow to a height of 1.8m, but on females does not normally exceed 0.9m.*

# LONGFIN PILOT WHALES

There are two species of Pilot Whales, the Longfin and the Shortfin. The Shortfin Pilot Whales are generally found in tropical and subtropical waters and are rarely observed around New Zealand. In contrast, the Longfin's range includes temperate to subantarctic waters, and they are the pilot whales which are seen at Kaikoura. Males have exceeded 7m in size although their average length is 5.5-6m. Females are usually less than 5m long, although some have reached sizes of up to 5.5m. As their name suggests, these whales have very long flippers which are up to 20% of their total body length.

The bodies of Pilot Whales are mostly black except for a whitish or pale gray anchor-shaped patch on their throat, which extends into a pale stripe along their belly. In the Southern Hemisphere, many Pilot Whales exhibit a white elongated teardrop patch behind their eye and a saddle patch behind their dorsal fin. Like other toothed whales, Pilot Whales use echolocation, and in addition they produce a large variety of complex sounds which make them great fun to listen to. It is believed that each individual whale has his own "signature" whistle, and it's also possible that, like Orca, groups have their own dialects. Longfins feed primarily on squid, but like their Sperm Whale cousins they also enjoy variety and add other fish species to their menu. They are deep divers who are known to reach depths of at least 600m, although they normally feed in shallower depths between 50 and 100m. Again, like their Sperm Whale cousins, Longfins are considered to be an offshore species rather than a coastal species. There is, however, a tendency for them to move closer inshore during summer months, perhaps as a result of the numbers of calves born at this time. Their gestation period is a long one (16 months), and while calves are born throughout the year, spring and summer births seem to predominate.

In Kaikoura, the majority of our sightings occur in summer, from mid-December until March. We see groups that range between 50 and 300 individuals, with our average group size ranging from 100 to 150. We normally see the Longfins quite a way offshore (15-25km), but in one instance we had a great time with a group of around 75 who were in close, swimming with Dusky Dolphins. The dolphins thought the Pilot Whales were great "toys" – they pretended the whales were boats, bowriding off their heads, playing leapfrog and jumping back and forth over their bodies. Bottlenose Dolphins frequently associate with Pilot Whales, but we had never before observed Dusky Dolphins with them. While we were wondering what had brought the Pilot Whales in so close to shore, we noticed a very tight group with females on the inside and an "escort" of males around them. Upon closer inspection, we saw what must have been a brand new calf. Its birth rings were evident and its dorsal fin was still completely folded over. There was one adult female very close to each side of the calf, and it barely went under the water's surface before taking each breath. It was amazing seeing this tiny creature that had just entered a new world and to observe the protection being given by the adults. We moved away in order not to disturb the new family and went on to enjoy the other whales and dolphins as they "played" together.

*The Pilot Whale is also known as the "Pothead" whale, a name reflecting the rounded head and bulbous forehead.*

*Pilot whales are one of the more socially cohesive whales, and are often sighted in large groups of several hundred with some reported sightings of over 1,000 individuals.*

# STRANDINGS

There are few sights more exhilarating than watching living whales and dolphins at home in their world, the sea. Conversely, there are few sights more heartbreaking than seeing whales or dolphins helplessly stranded in our world, the land. Whale strandings have long been a mystery to mankind. Why should these creatures from the ocean's depths seemingly "commit suicide" on our beaches. Single whale strandings are easier to understand for they are almost always the result of illness, injury, or death. The reasons for mass strandings of seemingly healthy animals, however, have been the subject of much conjecture. One common feature of mass strandings, however, is that they occur almost exclusively among Odontocetes, or whales which use sonar to help navigate. Another is that the majority of mass strandings occur in animals which are normally considered to be offshore species, and who travel in medium to large groups, such as Pilot and Sperm Whales.

Two major factors have been singled out as the probable causes of the majority of mass strandings in NZ waters.
1. Illness: In some stranded animals, parasite infections have been found which may have interfered with a whale's ability to echolocate or with its equilibrium. In other instances, a weak animal may be unable to continue swimming and gradually comes closer to the coast. No matter what the cause of the initial stranding, once whales become helpless they emit piercing distress calls and all the other whales in the group come to their aid, becoming stranded victims themselves.
2. Physical characteristics of the land: Areas which have sand spits or long fingers of land which jut into the sea seem to act as whale "traps". The whales obviously intend to travel around these projections in the open sea, but sometimes they end up inside the spits and become trapped. The sea bottom in these areas is usually gently sloping and sandy, characteristics which reduce or distort the effectiveness of a whale's sonar. Additionally, these areas are usually very tidal, and many strandings occur during periods of maximum tidal fluctuations.

When a few animals become stranded, the rest attempt assistance. In New Zealand, many techniques have been developed to aid stranded whales and dolphins, which have led to some very successful rescues. Following are essential things to keep in mind if you ever become involved in a whale or dolphin rescue:

**DO:**
1. Call for help immediately, giving the exact location, number of whales and species (if known), as well as the sea and tide conditions.
2. Make sure that the whale's blowhole is out of the water so it can breathe. Often the whales tip over as the tide recedes and it is essential to help them get upright as the tide moves in, otherwise they drown.
3. Keep the whales shaded and cool. Wet sheets work extremely well, but at the very least pour water over them to keep them from overheating.

4. Stay calm and talk to the whales. Once the tide comes in, it is important to keep the whales together and try to move them offshore as a group.

**DON'T:**
1. Never pour water down a whale's blowhole.
2. Don't try to move the whales by pulling on their flippers, dorsal fins, or flukes. Wait until there is sufficient water before trying to move them.
3. Don't separate mums and calves.

It's an almost indescribably good feeling to watch rescued whales swimming freely in the open sea once again. We may never be able to prevent strandings, but improved rescue techniques and the help of caring humans can save large numbers of whales.

*In January 1991 close to 300 Pilot Whales stranded inside Farewell Spit; 48 exhausting hours later rescuers had returned over 250 whales to sea. It was intensely emotional – both sadness and joy.*

*Whales become stuck in the sand or mud as tides recede and must be helped upright as the tide moves in.*

# DUSKY DOLPHINS

The Sperm Whales may be touted as the stars of Kaikoura, but it's often the Dusky Dolphins who steal the show! Even "hardened" whale watchers who proclaim that they have seen "hundreds of dolphins and just want to view whales", end up being totally captivated by the antics of the Duskies.

These charming dolphins are not very large; their maximum length is 2.1m, but they only average between 1.6m and 1.8m. Duskies are a Southern Hemisphere dolphin usually found in temperate waters and often observed in groups of hundreds. In Kaikoura, sightings of Dusky Dolphins occur throughout the year. From mid-late October through May Duskies are usually found close to shore from early morning to late afternoon. In winter, the dolphins tend to move offshore where they are sometimes sighted in very large groups. We've had a few incredible days with over 1,000 Duskies playing around our boats.

Reference or guide books used to proclaim that New Zealand Duskies give birth during the winter months. Kaikoura's Duskies obviously did not read those books! We have observed these dolphins for many years, and the vast majority of the births occur from early spring to early summer, with new calves also occasionally appearing in late summer. In the peak calving periods, mums and their new offspring often form nursery groups of 6 to 20 mums with their babies. These groups sometimes remain segregated from the larger pod of dolphins until the babies are a month or two old. One day we were privileged to encounter a brand new baby with its umbilical cord still visible, and its dorsal fin still folded over. The baby was being helped at the surface by the mother and another adult. It was already popping up and down on its own with quick little breaths, an amazing sight!

Dusky Dolphins sometimes "munch" throughout the day, but night is their peak feeding time. Their prey includes small schooling fish such as yellow-eyed mullet, kahawai, lantern fish, and small squid. They "fish" cooperatively much of the time using echolocation and a variety of whistles, squeaks, and squeals to communicate with

each other. They also use their bodies to help herd fish, by doing a series of leaps, lunges and tailslaps. Picture in your mind a large group of dolphins chasing a big school of fish with the dolphins near the front and on the sides creating a tremendous "thwack" as their bodies hit the water. The fish surrounded by "thwacks" on each side head for the middle where the majority of dolphins happily scoop them up.

I was lucky enough to be in the water one day to observe another feeding technique with the dolphins rounding up hundreds of kahawai into a tight ball. About 15 dolphins were swimming round and round their prey echolocating like mad. The kahawai were all tightly bunched at the surface, while the remaining dolphins swam leisurely under them and plucked out the fish of his or her choice. The dolphins would take turns with feeding dolphins joining the herding group and vice versa. I remained transfixed for about 45 minutes before my companions plucked me out of the water.

The Duskies are very gregarious and really seem to enjoy the contact with boats and people. They will sometimes play and "dolphin talk" with swimmers for hours, but there are other times when they are busy with their lives and only approach for a quick "hello". Their gregariousness extends to other whale and dolphin species, and we have seen them interacting with Common Dolphins, Bottlenose Dolphins, Pilot Whales, Sperm Whales, Southern Right Whale Dolphins, and even Orca.

*The exuberant behaviour of Dusky Dolphins has given them the well desrved title "The acrobats of the sea".*

*A Dusky mum with a young calf which displays fading birth rings on its body.*

*These leaping Dusky Dolphins are responding to an underwater "chorus" of sounds. This "chorus", which we've heard on other occasions, appears to be an announcement from fellow Duskies who have located a plentiful supply of fish in another area.*

*Duskies are smaller than Common Dolphins with a different colour pattern and a much shorter beak.*

# COMMON DOLPHINS

*"The dolphin is provided with a blowhole and lungs… and has been seen asleep with his nose above the water, and when asleep he snores. Not one is ever seen to be supplied with eggs, but directly with an embryo, just as in the case of mankind. Its period of gestation is 10 months and it brings forth its young in summer. The dolphin is provided with milk and suckles its young, which accompany it for a considerable period. In fact, the creature is remarkable for the strength of its parental affection. The young grow rapidly, being full grown at ten years of age. It lives for many years; some are known to have lived for more than twenty-five and some for thirty. The fact is, fishermen nick their tails sometimes and set them adrift again, and by this expedient their ages are ascertained."*

The Greek philosopher Aristotle wrote the above description of the Common Dolphin 2300 years ago. The description is amazingly accurate, and one of the first recognitions of dolphins as mammals. Common dolphins, as the name suggests, are one of the most abundant dolphins and occur worldwide in tropical and temperate waters. They reach maximum lengths of 2.5m, but average 2-2.3m. As Aristotle correctly observed, their gestation period is between 10 and 11 months and calves are born in late spring and summer. Common dolphins are very gregarious and have been seen travelling in groups of thousands, almost literally a "sea of dolphins". They are considered a "pelagic" or offshore species, normally travelling outside the 100 fathom contour line. They can dive to a depth of 280m and remain submerged for at least 5 minutes, and possibly longer.

Common Dolphins "visit" us most frequently in February and March, and they often spend their time in Kaikoura playing and feeding with our local Dusky Dolphins. In 1990 we noticed a Dusky Dolphin mum with a hybrid calf, half Dusky/half Common, and in the Marlborough Sounds in late summer we saw a dolphin that appeared to be a Common/Bottlenose mix, a 3m grey dolphin with the distinctive Common Dolphin golden blaze on its sides. It thus appears that interspecies "communication" may be taking place in New Zealand's waters!

*Common Dolphins have a distinctive tan or golden blaze on their flanks and a pronounced beak.*

# HECTOR'S DOLPHINS

The Hector's Dolphin was named after a New Zealand zoologist, Sir James Hector, who first described it in 1869. These dolphins are only found in New Zealand waters, with a primary range around the South Island, and a few localized sightings reported from the North Island. Hector's Dolphins are very coastal animals, and the majority of the population keeps within a few kilometres of the shore. They seem to have a preference for murky water and are often sighted near river mouths. They feed on small schooling fish such as yellow-eyed mullet, herring, small cod, kahawai, and also small squid. It's very interesting that the largest toothed cetacean (the Sperm Whale), and one of the smallest of toothed cetaceans (the Hector's Dolphin) only produce pulsed echolocation clicks as sounds. In the case of Hector's Dolphins these are very fast, high frequency clicks which are emitted at rates of 10-50 per second. The maximum click rate recorded by Dr Stephen Dawson has been over 1100 clicks per second!! There are no known whistles or other calls produced by these dolphins.

The Hector's Dolphin has gained a wide range of nicknames, the "Puffing Pig", the "Dumpling Dolphin", the "Mickey Mouse Dolphin", and the "New Zealand" or "Down Under Dolphin". The "Dumpling Dolphin" title refers rather disparagingly to its small rounded stature. At maximum lengths of 1.4m and weights of 48kg, the Hector's is the smallest marine dolphin in the world.

*The "Mickey Mouse Dolphin"! One of the Hector's most distinctive characteristics is its rounded dorsal fin. This uniquely-shaped dorsal makes it difficult to confuse Hector's with other dolphin species in New Zealand waters and has led to the nickname of "Mickey Mouse Dolphin".*

In recent years, research by New Zealand marine biologists Dr Alan Baker, Dr Stephen Dawson, and Dr Elizabeth Slooten has given us new and invaluable information about Hector's Dolphins. They are an endangered species with population estimates of only 7,000-8,000. Their life span is relatively short, with maximum ages of approximately 20 years. Females are sexually mature at between 7 and 9 years of age and only produce calves every 2-3 years at best, with their tiny 76cm calves generally born in late spring or early summer. Obviously the number of calves potentially produced in a lifetime is quite small. Hector's are normally found in small groups of 2-12 individuals and they appear to have fairly stable local populations. Occasionally you may see "gatherings" of 30-60 individuals, but these "get togethers" are the exception rather than the rule. Hector's Dolphins display a variety of behaviours, particularly during their "gatherings", and these include spy-hopping, chasing, leapfrogging (one dolphin jumping on top of anther), and leaping. In general however these dolphins are rather sedate and can appear to be quite shy, and sometimes seem to reserve their exuberant behaviour until you've run out of film or are looking the other way!

Unfortunately, their inshore habitat makes Hector's Dolphins very susceptible to entanglement in gill-nets set by fishermen. Many known deaths have already occurred, and even more frightening is the potential number of unreported deaths. Around Banks Peninsula near Christchurch the threat to the Hector's Dolphin has seen a sanctuary established which prohibits both amateur and commercial gill-netting during the summer months. This is a good start, but it is only a start. Given the Hector's small population and slow reproductive rate, any gill-net death is one death too many. It is, after all, the "New Zealand Dolphin", and we must look at ways of protecting and preserving the species throughout its very limited range.

In Kaikoura we are fortunate that few known deaths have been reported among our local populations. One of our established populations lives approximately 20km north of the Kaikoura peninsula, another one lives around 20km south of the peninsula.

Additionally, there are other "pockets" of Hector's Dolphins which are sighted at certain locations. We have, on occasion, enjoyed groups of between 30-40 individuals and one astounding day counted up to 60 dolphins; but our sightings are usually of the typically small numbers ranging from 2-15 dolphins. "Our" Hector's are with us all year round and on the occasional days when we miss seeing them, we know they haven't travelled far. Even though most births take place during spring and summer, we had one group that established it's own timetable, and for three years a new calf appeared in late autumn. A "Bonus" baby! We are extremely fortunate to have this rare and "adorable" dolphin living within our natural marineland. It's always astounding to think that we have both one of the largest and the smallest cetaceans, the Sperm Whale and the Hector's Dolphin, often residing within a ten minute boat ride of each other, and even more astounding when you contemplate the other marine species that are sometimes observed between them.

*Hector's are relatively slow swimmers, and only occasionally exhibit short bursts of speed exceeding 10 knots.*

*1.4m Hector's – 6m boat.*

# SOUTHERN RIGHT WHALE DOLPHINS

The Southern Right Whale Dolphin's most outstanding characteristic is its lack of a dorsal fin, prompting their descriptive scientific name "Lissodelphis", from the Greek "lisso" smooth, and "delphis" dolphin. There is a Northern and Southern Right Whale Dolphin with the northern species being somewhat larger and exhibiting a darker overall body colour.

Very little is known of this pelagic dolphin. The average length of the Southern Right Whale Dolphin appears to be 1.8m with maximum lengths of approximately 2.4m. They have a circumpolar range with the majority of sightings occurring in subantarctic waters. Around New Zealand, they are generally observed outside the 100 fathom contour. Stomach contents from stranded individuals indicate their main prey is squid, although remnants of a few other fish species have also been found. They are normally a shy species and do not often approach boats, but occasionally they will bow-ride for a short spell.

In Kaikoura we have sporadic, exciting glimpses of these dolphins, usually some distance from the land. About half the time they are sighted in the company of Dusky Dolphins, and they are certainly one of the most unusual cetaceans that we encounter off the Kaikoura coast.

*Like their distant relation, the Right Whale, Southern Right Whale Dolphins lack a dorsal fin, making their identification at sea an easy one.*

*Southern Right Whale Dolphins have been described as "dolphins who travel in a series of bouncing leaps", certainly an apt simile! They are extremely fast swimmers reaching speeds of at least 25 km/hr. (On occasions we've tried to keep up with them...we lost!) Note their "yin yang" colour pattern.*

# BOTTLENOSE DOLPHINS

Often when people hear the word "dolphin" the image of the Bottlenose leaps into mind, probably because many of us once saw "Flipper", the TV dolphin that was almost human. As well, many of the captive dolphins in marinelands are Bottlenose. The Bottlenose is also the dolphin which most frequently seeks association with people, and there are many stories worldwide of Bottlenose who develop close relationships with humans.

Bottlenose normally have a dark grey body, shading to lighter grey on the flanks and belly, and they are often heavily scarred. Their gestation period is 12 months, with calves normally born in summer. They are a gregarious dolphin, both with their own kind and with other species, often associating and even mating with other dolphins. They often travel with Pilot Whales, Right Whales, and Humpbacks. Some scientists feel there are two ecotypes of Bottlenose, with dolphins which are found inshore averaging 2.8-3m and much larger offshore dolphins reaching lengths of 4m.

In Kaikoura we seldom observe Bottlenose, and when we do they are almost always in the company of Pilot Whales. Interestingly, they do not often approach our boats, but instead go hurtling past at "a mile a minute" giving us flashing glimpses of their large grey bodies. And they are large. The Bottlenose we see appear to average at least 3.5m. Our sightings often occur 15-20km offshore and usually during summer months. Bottlenose are great copycats, and there are some fascinating stories of their mimicry of human behaviour. One such story is told of a human who blew a puff of cigarette smoke at the window of a viewing tank. The young dolphin on the other side of the window promptly went over to its mother and returned with a mouthful of milk which it "blew" back at the human. Another example of interspecies communication?

*New Zealand has had it's share of friendly dolphins. "Maui" delighted swimmers and observers at Kaikoura for over three years. She especially loved the "seaweed game". Maui now resides in the Marlborough Sounds where she is known as "Woody".*

# MYSTICETI: THE BALEEN WHALES

While the first surviving group of whales, the Odontoceti, retained their teeth, the second group developed baleen. Baleen plates hang closely together from the roof of the mouth and are used for feeding. Constructed of material similar to our fingernails, each baleen plate has a bristly fringed area along one side which faces inward. Each baleen whale's mouth contains hundreds of baleen plates.

Baleen differs between species. Humpback Whales, for example, have very coarse baleen which reaches lengths of up to 60cm (24in); Right Whales have very fine bristles which can reach lengths of up to 2.8m (9.2ft)! Some Baleen Whales feed by taking in huge mouthfuls of water containing prey, such as krill or small schooling fish like herring or capelin, and then forcing the water out of their mouths, utilizing their baleen as a sieve to trap the remaining food. Right Whales are "skimmers" who swim along with their mouths partially open, allowing the water to flow out the sides while the food becomes entangled in the baleen. Baleen whales do the majority of their feeding during the summer months, migrating into cold Arctic or Subantarctic waters where long daylight hours create conditions which promote the proliferation of krill and other food sources. During autumn and winter many baleen whales move into warmer waters to breed and give birth.

There are 13 species of baleen whales, including the mighty Blue Whale. Blue Whales have reached lengths exceeding 33m (100ft) and weights exceeding 150 tonnes. It is amazing that possibly the largest creature ever to have inhabited the earth feeds on some of the smallest creatures in the sea. It is also interesting to note that among the baleen whales the female is larger than the male in the majority of species. The reverse is true of the toothed whales and dolphins. This phenomenon may be due to the long migrations undertaken by the baleen whales during which little or no feeding occurs. The females would therefore need more mass in order to sustain both themselves and their young nursing calves. Baleen whales are not thought to have an echolocation system, and the sounds they produce range from the haunting and complex songs of the Humpback Whales to the low booming call of the Blue Whale.

A number of species of baleen whales were formerly a common sight off the Kaikoura coast, especially Humpbacks and Southern Rights. Sadly, however, the efficiency of the whalers virtually eliminated these species in New Zealand waters, and today they are rare visitors. Nonetheless, at least seven different species of baleen whales have been seen off Kaikoura in recent years.

*Lunging Humpback Whale.*

*Baleen whales all exhibit two external nostrils or blowholes.*
*Roger Sutherland – photo*

# HUMPBACK WHALES

Humpback Whales were once one of the most abundant large whales worldwide. In the Southern Hemisphere it is believed that over 100,000 Humpbacks once roamed the seas. 1996 estimates put the world population at about 12,000-15,000, a sad indictment of man's short-sightedness and greed. The humpback group which migrates from Tonga through New Zealand waters and into the Antarctic is tentatively numbered at 300-400, and sightings around New Zealand in recent years have been minimal.

Humpback whales are large, robust creatures with females reaching a maximum length of up to 16m and males, slightly smaller, up to 15m They weigh approximately one tonne per foot. A 16m female could weigh close to 50 tonnes… that's a lot of woman! They are relatively slow swimmers averaging 6-12 km/hr. Even so, they undertake long migrations, travelling from their winter breeding and calving grounds in warm tropical waters to their summer feeding grounds in the icy polar seas. Their diet consists of krill and small schooling fish such as herring. Humpbacks exhibit some "creative" feeding methods, including the "Bubble Net" technique where the whale encircles its prey from beneath the surface blowing air from its blowhole to form a "net" of bubbles. The whale then surfaces through the middle of the "net" with its mouth wide open. As the whale take in huge mouthfuls of water and food its pleated throat grooves expand somewhat like a pelican's pouch. The whale then contracts its throat and uses its large tongue to help expel the water. The "goodies" are all left behind, trapped in the 270-400 pairs of coarse baleen plates which hang from its mouth.

Female Humpbacks have a gestation period of approximately 12 months and normally calve at intervals of two years or more. The calves are weaned after approximately ten to twelve months of nursing. The breeding and calving occur in warm, tropical areas where the males engage in ritualistic courtship displays which include lunging, chasing, jaw clapping, and head butting. It's quite a spectacle to watch a female being pursued by ardent suitors all competing for her attention. Another method of wooing the females appears to be the "Humpback Serenade" where the male

*A Humpback calf spyhops displaying its head which is covered with fleshy protrusions.*

dives 10-20m underwater, and often hanging with his head down, emits haunting and complex songs. These songs may last from 10-30 minutes, and sometimes go on for hours, with the male surfacing to catch his breath and then returning to the depths to sing again. During a breeding season, all the whales sing basically the same song although individuals add their own signatures.

In New Zealand, whalers killed 200 Humpbacks in 1960, but in 1963 they could only find 9. However, the tide may be turning for the Humpback and populations throughout the world are increasing. In 2001, 7 whales were sighted in Kaikoura and a similar number are observed each year. Seven whales is a pitiful count compared to earlier days. We can only hope that each year will witness an increasing number of Humpbacks travelling through Kaikoura during their annual migrations.

*A diving Humpback displays the underside of its tail flukes. Each fluke exhibits a different scalloping pattern along the edge as well as pigmentation which is unique to each whale... somewhat like a human's fingerprints.*

*A breaching Humpback displays its pleated throat grooves. The number of grooves ranges from 20-35.*

# OTHER OCCASIONAL SIGHTINGS

Humpback Whales belong to a family called the "Rorqual Whales" that share the common characteristic of ventral throat grooves which expand when feeding. In addition to Humpbacks four other species or Rorquals have occasionally been observed in the Kaikoura area.

Fin Whales (below) are magnificent creatures, reaching 24m and only exceeded in size by the mighty Blue Whales. Their back is ridged from the dorsal fin to the tail flukes, giving them the name "Razorback Whales". Another common name used by the whalers was "Finback", derived from the pointed dorsal fin which usually appears after the blow. Like Sei Whales they were heavily exploited in the Southern Oceans in the 1950s and early 1960s after the decline of Blue Whales.

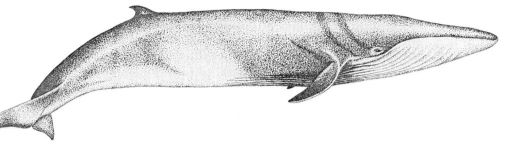

Minke whales (above) are the smallest of the Rorquals (9-10m). They have a narrow pointed snout and sharp dorsal fin which appears simultaneously with their blow. Their habit of appearing for a couple of blows and then vanishing completely has given them the affectionate nickname of "Slinky Minkes".

The Blue Whale (below) is the largest whale and the largest creature that has ever existed on this planet. One Blue Whale measured 33m and weighed over 190 tonnes. In the Southern Hemisphere over 95% of these whales were killed by whaling. In 1994, the IWC estimated the Southern Hemisphere population to be "a maximum of 1,000", which gives them the distinction of being one of the most endangered whale species. In recent years there have been sporadic sightings of Blue Whales around New Zealand and we are occasionally fortunate enough to experience a glimpse of this mighty leviathan as it cruises through Kaikoura.

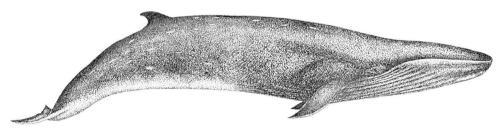

Sei Whales (above) are much bigger than Minkes (21m for females, 17.7m for males) and their head is more rounded with a single prominent rostral ridge running from the blowhole to the tip of the snout. When surfacing, the head, blowhole, back and dorsal fin appear almost simultaneously. Sei Whales are the "greyhounds" of the baleen whales and reach speeds of up to 30km/hr for short bursts.

*Drawings by Conrad Field*

*Mark Carwardine photo*

# SOUTHERN RIGHT WHALE

If the fate of the Humpback is a sad testament to human exploitation, the fate of the Right Whale is a tragic one. Once plentiful, both the Northern and Southern Right Whales suffered catastrophic depletion at the hands of whalers and were on the verge of extinction by the time they were protected in 1936. The Southern Right was extensively hunted from the early 1800s and was considered the "right" whale to catch, as its slow speed and coastal habitation during migration, breeding, and calving made it an easy prey for whalers. Additionally, the Right Whale provided a high oil yield, long and valuable baleen plates, and tended to stay afloat after it had been killed.

Right Whales are very robust creatures weighing up to 100 tonnes, with females reaching maximum lengths of 17-18m and males being slightly smaller. One of their most distinctive features is their complete lack of a dorsal fin; another is a very large head, which is up to 1/4 of their overall body length and covered with yellowish lumps or callosities. The largest growth of these callosities is located on the rostrum in front of the double blowhole, and is called the "bonnet". The arrangement of the callosities differs for each whale and is used by researchers to identify individuals. The Right Whale's blowholes are widely divergent so that their blow appears as a double spout. The Right Whale is also one of the few remaining whales which occasionally retains hair, sprouting on its chin and jaw.

Unlike the Humpback, the Right Whale has no ventral throat grooves and catches its food by swimming with its huge mouth agape, allowing the water to flow out the sides and leaving the krill and copepods entangled in the fine bristles of the baleen. Their long baleen plates grow to lengths of 2.8m. Southern Right Whales sometimes feed in the polar waters of the Southern Ocean but do not appear to make the extensive migrations that are undertaken by most baleen whales.

The New Zealand population of Southern Right Whales is estimated to be between 500-800. The majority of these whales are observed at two subantarctic islands, Campbell and Auckland Islands. Researchers at the Auckland Islands have reported an annual average of 20 new calves during the last three seasons ...a good sign! The sightings of Southern Right Whales around mainland New Zealand however are still so rare their numbers are considered to be "scientifically zero"... a poignant statement about a whale that was considered to be "common as mud" by early settlers.

Like the Humpbacks, we can only hope that protection did not arrive too late, and that future years will witness a return of these magnificent whales to the New Zealand coast.

*Winter in an area where the blows of Southern Right Whales were once common. Today we view only an empty sea.*

A Southern Right Whale displays its callosities which are raised patches of rough skin. The callosities form a unique pattern on the head of each Right Whale.

# CONSERVATION

Until the last few hundred years the oceans of the world teemed with life in all shapes and sizes. But from the late 1700s the large whales, the mighty leviathans of the deep, have been hunted relentlessly, some of them to the brink of extinction. As vessels and fishing technology have improved the harvesting of the seas has intensified, and virtually every developed country has had some share in the destruction of whales and the over-exploitation of the seas in general. As the oceans have become systematically depleted of life "save the whales" has become a major rallying cry, focusing attention not only on the fate of whales and other marine mammals, but also on the way human beings abuse other aspects of the natural order of things, of which we too are an inseparable part.

Every year over a million small whales, dolphins, and porpoises, continue to die!! This is a staggering figure, particularly in a time when cetaceans have become largely "protected". "How can this happen?" you may ask. Well, here's how!

• Deliberate murder: Tens of thousands of dolphins and small whales are deliberately killed for food, oil, sport, fertiliser, chicken feed, aphrodisiacs, and crab bait.

• "Incidental" or "accidental" catches: In the Eastern Tropical Pacific yellowfin tuna are often found swimming under schools of dolphins. Purse-seine fishing, which began in the late 1950s, has been responsible for millions of dolphins dying, trapped in nets that are deliberately set around dolphin schools in order to catch the tuna swimming beneath them. The good news is that after many years of lobbying and fighting, environmentalists are finally making inroads into saving the dolphins, with consumer pressure forcing some major United States tuna companies to stop buying tuna which has been caught by methods that lead to the inevitable deaths of dolphins.

• Oceanic Driftnets: In recent years many people have become aware of the disastrous result of driftnetting. These "walls of death" have been laid over great expanses of ocean, and show no discrimination as to what they catch and kill. It is no exaggeration to say that their continued use will result in irreparable damage to marine ecosystems over vast areas of the world's seas.

• Gill Nets: Invisible killers, modern gill nets are constructed of nylon line knotted together to form a mesh net. The sonar of whales and dolphins is usually incapable of discerning this fine mesh, turning the nets into invisible traps. Coastal populations of small dolphin and porpoises are the most susceptible to being caught and killed in this type of net. It is estimated that up to half a million dolphins and porpoise die annually in gill nets, but many of these deaths go unreported and their bodies are "quietly" dumped back into the sea. Even more tragic is the fact that these same dolphins and porpoises usually come from relatively small populations.

*Hector's Dolphin entangled in a gill net.   Steve Dawson -photo*

• Pollution: In Canada there is a white whale, the Beluga, which spends much of its life in the St Lawrence River. Man's pollution has made this one of the most contaminated mammals in the world. Under Canadian law, the corpses of Beluga Whales are treated as toxic waste! Man continues to use the oceans, the rivers, and the lakes as a dumping ground for waste products and poisonous, toxic materials that we don't know "what else to do with". It has been "out of sight, out of mind"… but is now coming home to us in the flesh of the fish we eat, in the sea, rivers and lakes we used to be able to swim in, and in the contaminated water that we have to drink. Our oceans cannot sustain the oil spills, the industrial waste, the plastic, and all the other pollutants that we dump into them with little thought for the long-term consequences. Perhaps man should wake up!

I would urge everyone to read Michael Donoghue and Annie Wheeler's book "Save the Dolphins". The deaths of over a million small whales and dolphins a year is tragic. More so is man's lack of respect for, and awareness of, the health of his total environment. It's time we stopped viewing the world through our own selfish eyes and begin to recognize that all living things are connected, and where nature has turned on man, man has first usually altered the course of natural things. For many species on both land and sea, it may already be too late. The tide must be turned in order to preserve some sense of natural balance, on our planet and in our lives. If not, our own obituary may now be being written in the fate of the whales and other endangered creatures who share our planet.

*The Hooker's Sea Lion (left) and Yellow-eyed Penguins (above) are, like many whales and dolphins, endangered species. Both are endemic to New Zealand waters and occasionally sighted at Kaikoura.*

# THE ETIQUETTE AND THE ART

## The Etiquette

Man often has an egotistical approach to nature and to nature's creatures. When we interact with our "fellow" humans we try to display good manners and proper behaviour. We would think it incredibly rude to arrive unannounced at a stranger's home, to drive quickly and loudly around the yard tossing out bits of garbage, and then to burst through the door without an invitation. Our terrible manners would continue if once inside we started touching and hugging everyone in sight, and ran around trying to observe bathroom and mating rituals. Why then do we break the rules of proper etiquette when visiting animals in their homes, and exhibit some or all of the above rude behaviour?.

When the animals react to our disrespect with aggression or retreat, we go away from the encounter feeling cheated and disappointed. What we need to remember is that all relationships take time; they need nurturing, familiarity, and a chance to build mutual trust. These simple rules apply to interspecies, as well as interhuman, relationships.

The animal kingdom is just that... their kingdom! When we visit animals in the wild, we enter their world, their home, and we usually arrive as strangers. If we don't have the time in our "busy" lives to get acquainted, then it is doubly important that we pay our respects with a minimum of intrusion.

When we buy a ticket to go whale watching or take part in any experience involving nature, it is man who is selling us that ticket, not the animals. The creatures that we encounter must have enough space to continue their daily lives without feeling threatened or harassed. It is a privilege for us to enter their world, not our right. If we value the time we spend with nature, we will utilize it wisely to observe and learn about the other life forms that we are fortunate to view. Most importantly, when we depart, we should leave behind a positive memory of our shared encounter.

## The Art

Man visits nature carrying the heavy baggage of civilization... suntan lotion, insect repellent, cameras, film, tripods, videos, tape recorders and field guides. We then go home and try to store our memories in a photo album or on video tape. But how can you store the essence of a whale in plastic cellophane between cardboard covers? The filmed images are not necessarily bad, but they tend to be only a superficial portrayal. And the field guides may be useful but are a bit like a name tag on a person's chest – they give us the facts but little of the soul. The next time you visit nature, try to immerse yourself into your surroundings before placing the black box in front of your vision. Spend some time watching, listening, smelling, and, if appropriate, touching. Feel the rough texture of the tree's bark; close your eyes and listen to the whale's blow, visualizing the cloud of white mist which emerges with each breath; watch the birds gliding effortlessly on the wind's currents; enjoy the interaction of the creatures you are observing as they work and play...capture the essence in your heart before you try to capture it on film. If you take the time to "feel" what you are seeing, the images you view later will be more than a piece of printed paper or a picture on a screen. Your memories will be richer and so too will your life.

# CHIEF SEATTLE

In 1854 the "Great White Chief" in Washington made an offer for a large amount of Indian land. Chief Seattle's reply to that offer has been described as one of the most beautiful and profound statements ever made on the environment.

"Every part of this earth is sacred to my people. If we sell you this land, you must remember that it is sacred and you must teach your children that it is sacred and that each ghostly reflection in the clear waters of the lakes tells of events and memories in the life of my people. We know that the white man does not understand our ways. One portion of the land is the same to him as the next, for he is a stranger who comes in the night and takes from the land whatever he needs. The earth is not his brother but his enemy and when he has conquered it, he moves on. He treats his mother, the earth, and his brother, the sky, as things to be bought, sold, and plundered like sheep or bright beads. His appetite will devour the earth and leave behind only a desert.

I do not know. Our ways are different than your ways. But perhaps it is because I am a savage and do not understand.

The air is precious to the red man, for all things share the same breath - the beast, the tree, the man, they all share the same breath. The white man does not seem to notice the air he breathes. Like a man dying for many days, he is numb to the stench. But if we sell you our land, you must remember that the air is precious to us, that the air shares its spirit with all the life it supports.

So we will consider your offer to buy our land. If we decide to accept, I will make one condition; The white man must treat the beasts of this land as his brothers. I am a savage and I do not understand any other way. What is man without the beasts? If all the beasts were gone, man would die from a great loneliness of spirit. Whatever happens to the beasts, soon happens to man... all things are connected.

Teach your children what we have taught our children, that the earth is our mother. Whatever befalls the earth, befalls the sons of the earth. If men spit upon the ground they spit upon themselves. This we know; the earth does not belong to man: man belongs to the earth. All things are connected. Man did not weave the web of life; he is merely a strand in it. Whatever he does to the web, he does to himself.

We may be brothers after all, we shall see. One thing we know that the white man may one day discover - our God is the same God. You may think you own him as you wish to own our land; but you cannot. To harm the earth is to heap contempt on its Creator. The whites too shall pass; perhaps sooner than all other tribes. Contaminate your bed, and you will one night suffocate in your own waste. Your destiny is a mystery to us for we do not understand when the buffalo are all slaughtered, the wild horse are tamed, the secret corners of the forest heavy with the scent of many men, and the view of the hills blotted by talking wires.

Where is the thicket? Gone
Where is the eagle? Gone
The end of living and the beginning of survival."

# FURTHER READING

I hope that you have enjoyed this introduction to the whales and dolphins of Kaikoura, that it has made some impact on your thinking, and perhaps left you curious to learn a bit more about some of our fellow mammals that live in the sea. Below is a short list of reference books which may be of interest. Remember when reading these books that the study of whales and dolphins is a living science, with changing ideas and new information continually coming to light.

Baker, Alan N. Whales and Dolphins of New Zealand and Australia, Victoria University Press, Wellington, 1999

Carwardine, Mark, Collins Wild Guide to Whales and Dolphins, Harper Collins Publishers, London, 2006

Doak, Wade, Encounters with Whales and Dolphins, Hodder and Stoughton, NZ 1988

Donaghue, Michael and Wheeler, Annie, Save the Dolphins, David Bateman, Auckland 1990

Folkens, Pieter A. Guide to Marine Mammals of the World, Alfred Knopf, London 2002

Payne, Roger, Among Whales, Scribner, New York, 1995

Simmonds, Mark P. Whales and Dolphins of the World, New Holland Press, 2004

# PUBLISHING INFORMATION

TEXT: Barbara Todd.

PHOTOGRAPHY: Barbara Todd unless otherwise indicated.

PRINTING: Spectrum Print, Christchurch, New Zealand.

PUBLISHED BY: Barbara Todd, Nature Down Under, Box 249, Nelson, New Zealand.

ISBN 978-0-473-13098-5

© Barbara Todd, 1991.

Reprinted 1992, 1995, 1997, 1999, 2002

Revised edition 2007